Beginning Maps

The activities in this book help children to understand:
- a map shows where things are located
- a map can help us find places

What Is a Map?

A map shows where things are.

This map shows what I see when I look down from the sky.

Can you find these places on my map?

Put an X on the 🏠.

Put a circle around the 🪑.

How many 🌳 did you find?

Color them green.

2

Parents: Explain to your child that pictures on a map look different than real objects. Then have your child match the map symbols to pictures of a "real" thing.

Pictures on a Map

maps

Match the pictures.

3

Lunch Time

This is a map of a table.
It is set for lunch.

EMC 4130

Parents: Your child is to cut out the pictures on page 23 and paste them on this table just as they are placed on the map.

Set the Table

Look at the map on page 4.

Color.   Cut.   Paste.

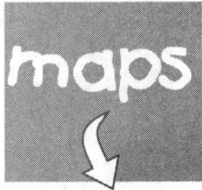

maps

Play Time Map

This is a map of the rug in my room.

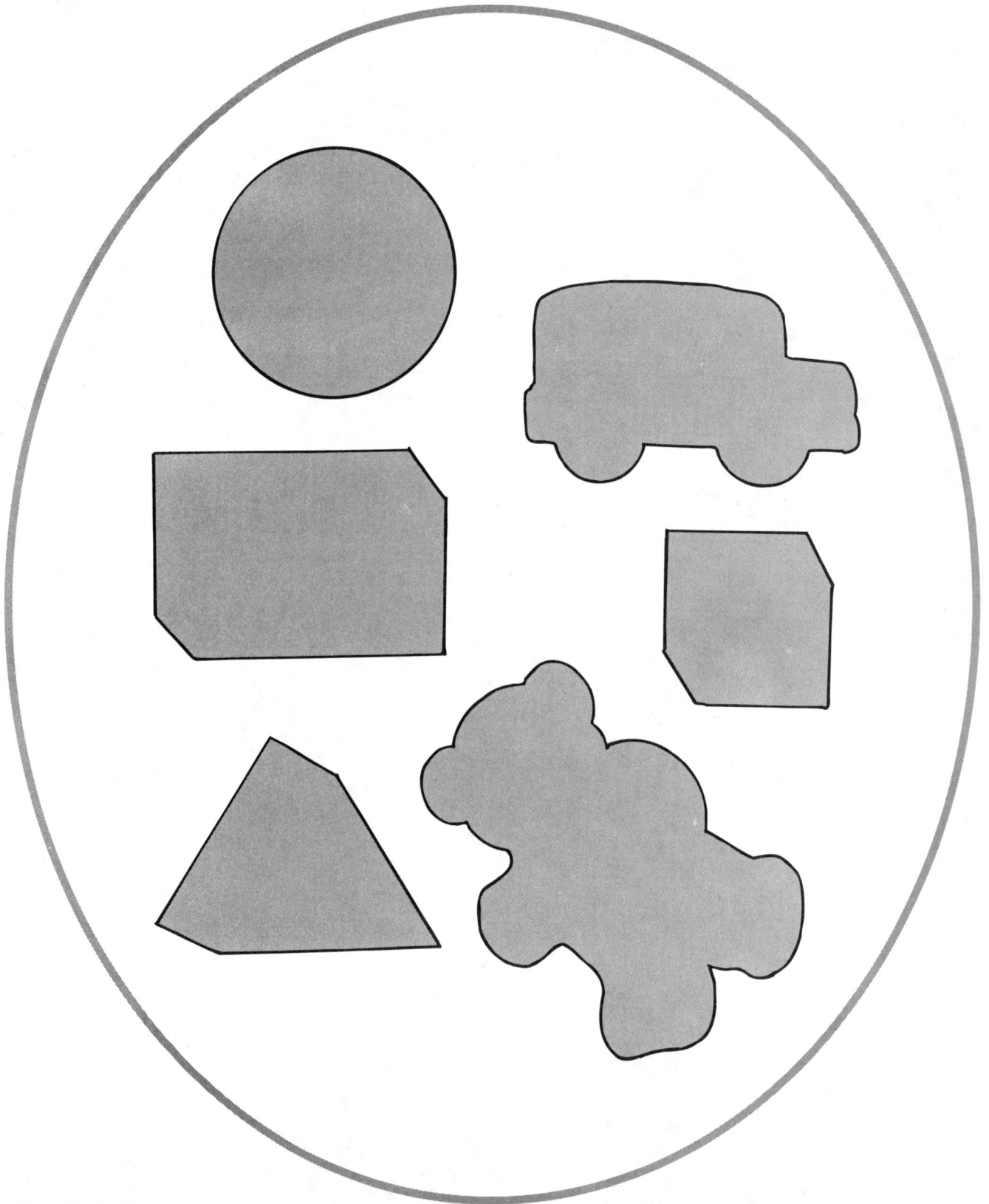

Parents: Your child is to cut out the pictures on page 25 and paste them on this rug just as they are placed on the map.

My Play Time Rug

Look at the map on page 6.

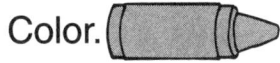

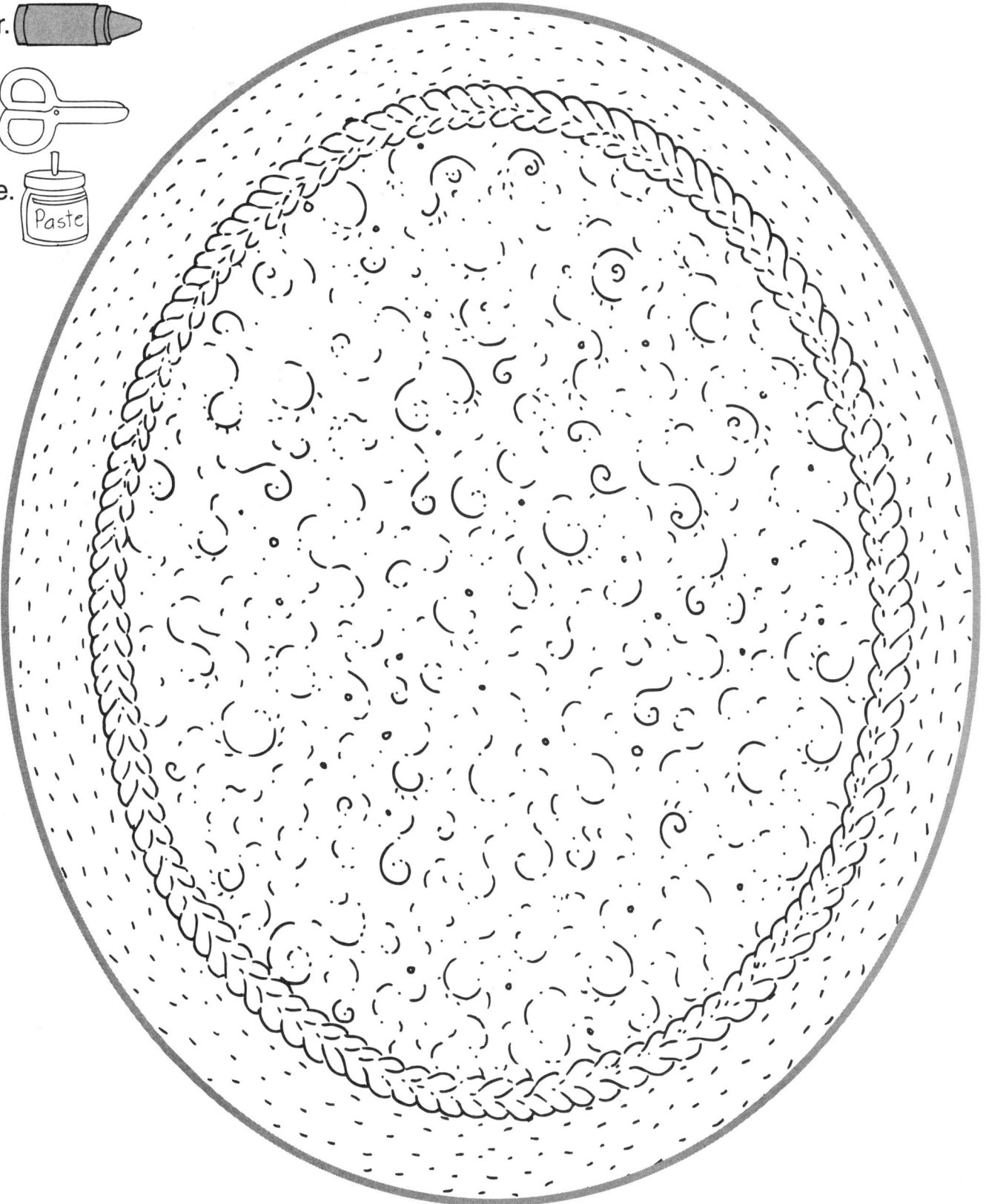

Color.

Cut.

Paste.

Bedtime

This is my bedroom.

I made this map of my bedroom.

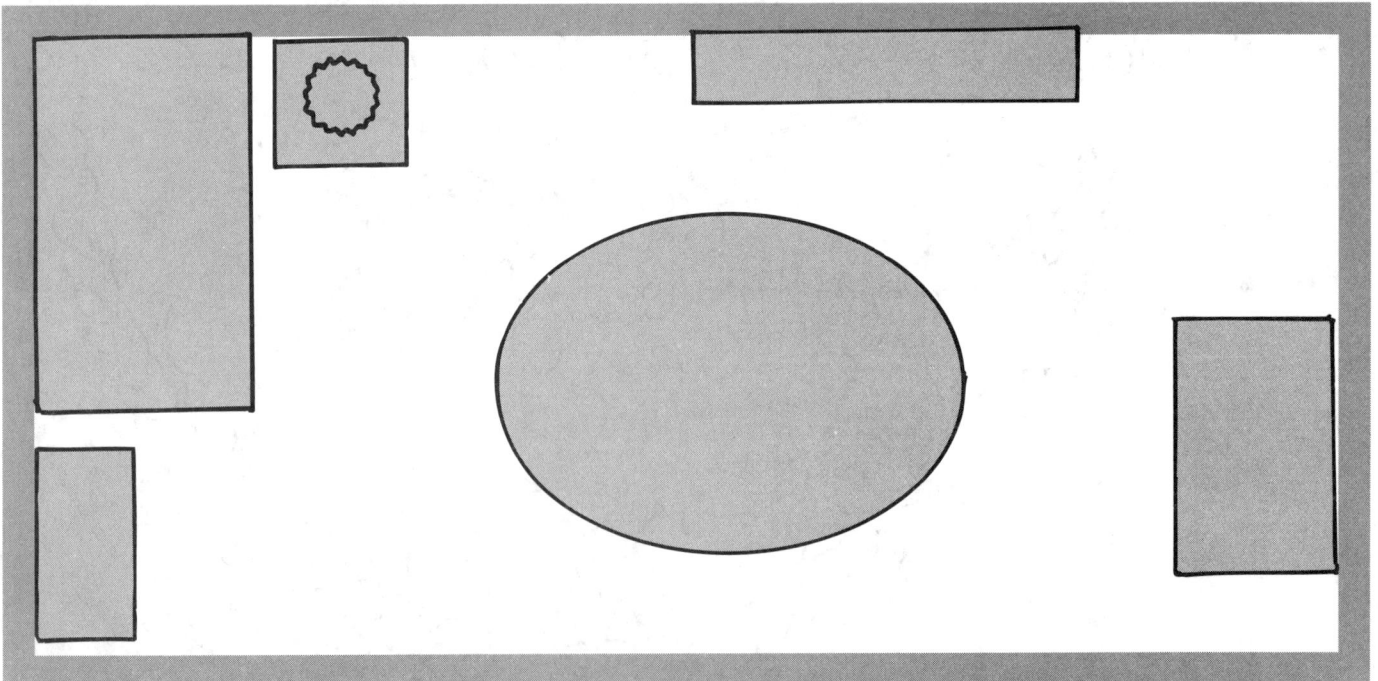

What does your bedroom look like?

Can you make a map to show me?

EMC 4130

Parents: Help your child mark the places where windows and doors belong on the box below. Have your child use the symbols for bedroom furniture on page 27 to make a map of his/her own bedroom.

Map Your Bedroom

Go to page 27.

Color. 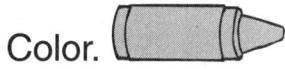 Cut. 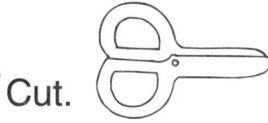 Paste.

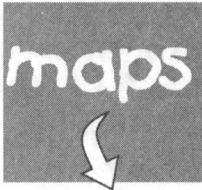

Parents: Your child is to draw a line from place to place to see where Rowdy goes in the yard. Help your child read the directions.

In the Backyard

This is Rowdy.

He likes to play in the backyard.

Use the map on page 11 to help you find Rowdy's favorite places.

Get a Red

Make a line where goes.

Start at .

Go to the .

Go from the to the .

Rowdy likes to chase the .

Go from the to the .

Rowdy likes to dig for here.

Go from the to the .

Rowdy wants to take a nap.

EMC 4130

Rowdy's Backyard

Parents: Help your child learn the four directions. Have your child start at the star in the center of the page. Have him/her trace the line up the arrow to north. Say "This line points north." Repeat this with each of the directions. The next few page practice the four directions.

Which Way Is It?

There are directions on a map.

These directions help you find places.

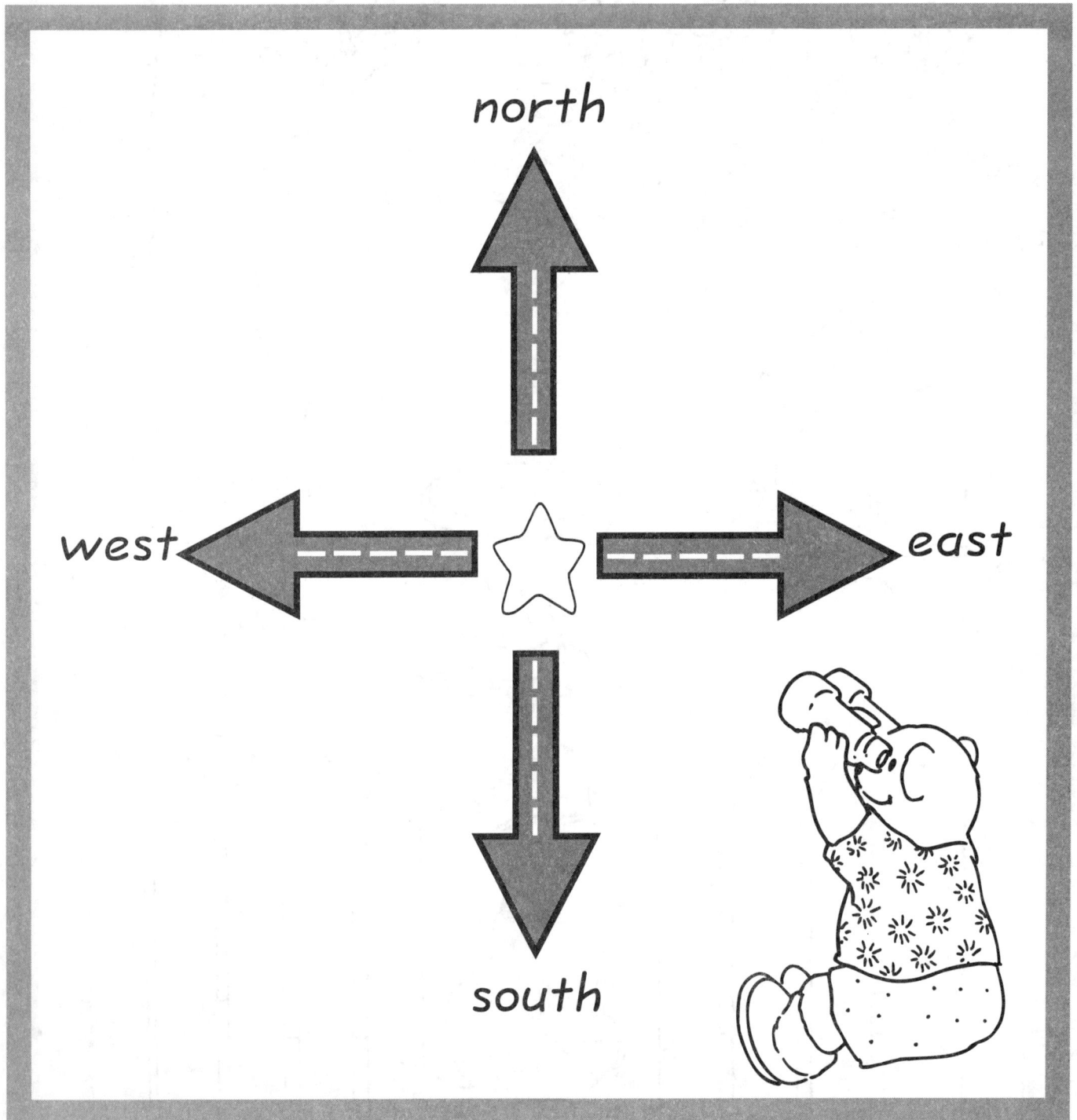

north

west ← → east

south

EMC 4130

maps

Find the Picture

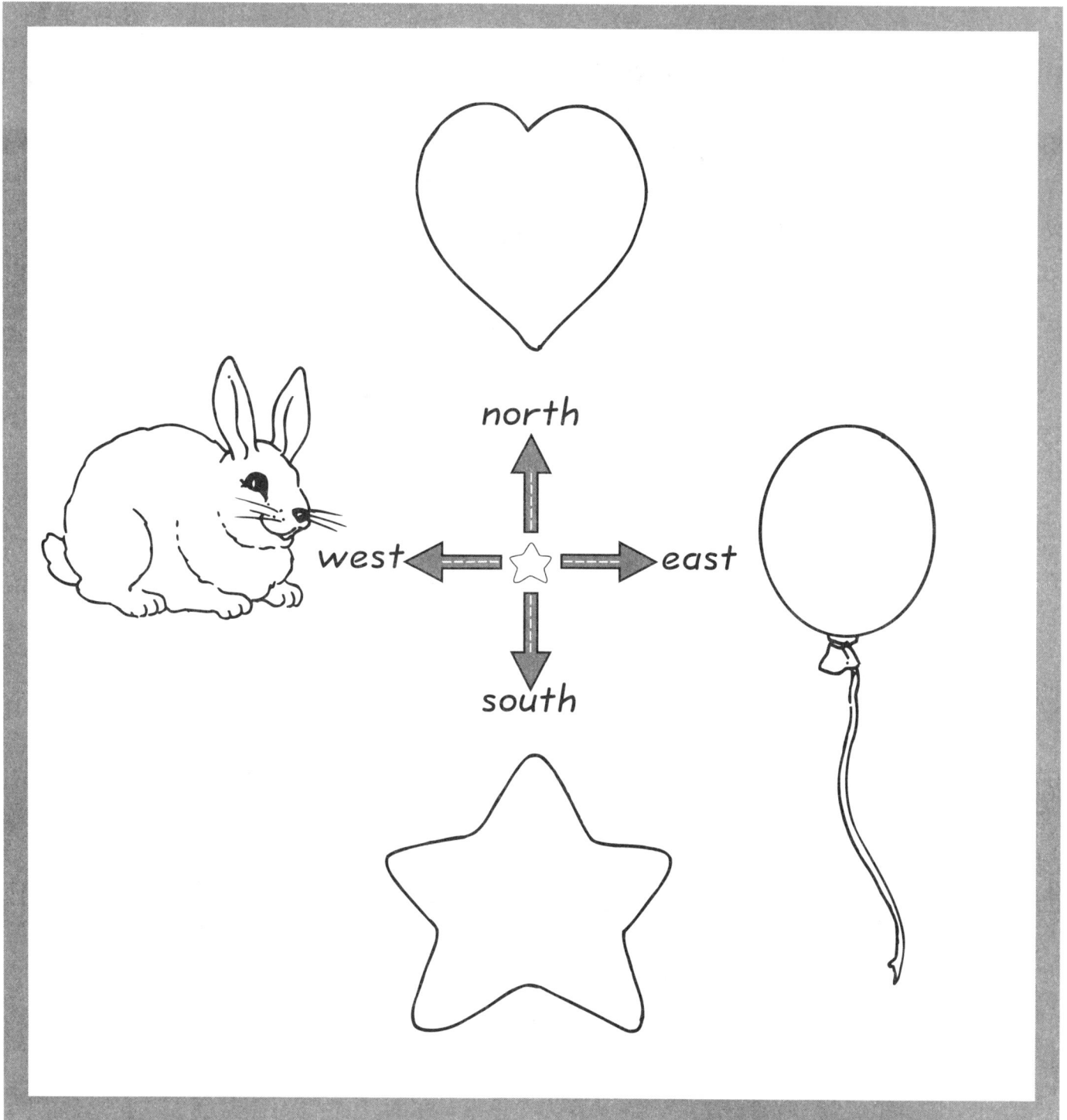

north

west ☆ east

south

What is **north**?	What is **west**?	What is **south**?	What is **east**?
Color it red.	Color it brown.	Color it yellow.	Color it blue.

Little Boy Blue

Go to page 29.
Color. 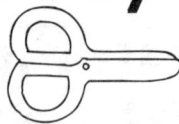 Cut. Paste.

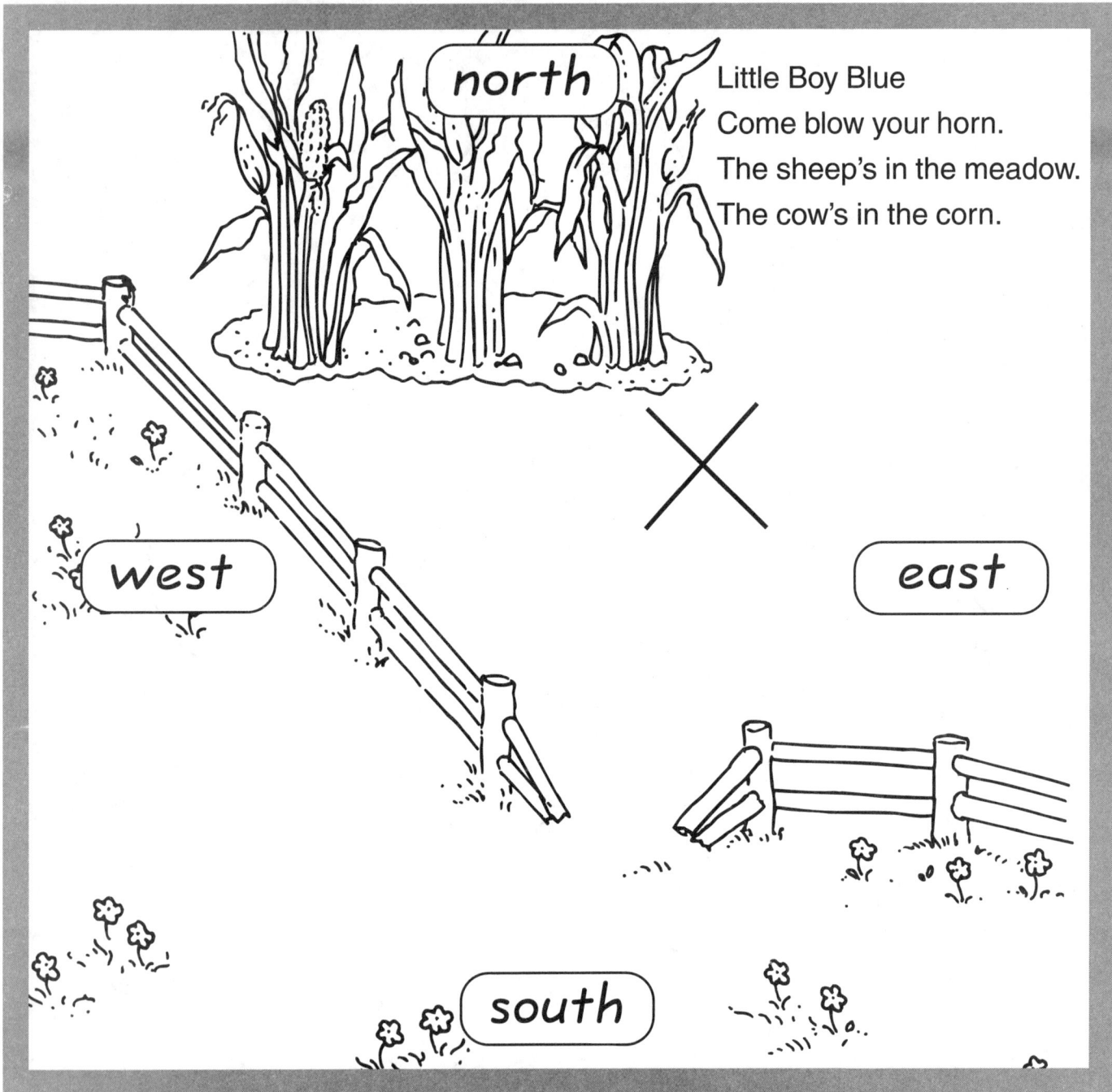

north

Little Boy Blue
Come blow your horn.
The sheep's in the meadow.
The cow's in the corn.

west

east

south

1. Paste ⬡ on the X.

2. Paste 🐑 west.

3. Paste 🐄 north.

4. Paste 👦 east.

EMC 4130

Little Bo-Peep

Go to page 31.

Color. 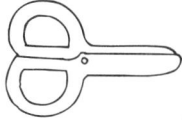 Cut. Paste

maps

Little Bo-Peep has lost her sheep.
Use the map to help her find them.

north

west ⟵ ☆ ⟶ east

south

1. Paste 2 🐑 west.

2. Paste 1 🐑 east.

3. Paste 3 🐑 south.

4. Paste 1 🐑 north.

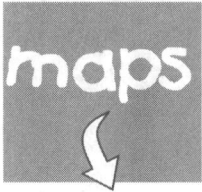

At the Petting Zoo

The zoo is a big place.

Look at the map on page 17 to help you find the animals.

Color:

What is north of the [calf] ?

Color it **brown** .

What is south of the [duck] ?

Color it **green** .

What is east of the [calf] ?

Color it **yellow** .

What is north of the [duck] ?

Color it **black** .

What is west of the [turtle] ?

Color it **pink** .

What is north of the [pig] ?

Color it **brown** .

Color the [calf] **brown** and **white** .

EMC 4130

Petting Zoo Map

Petting Zoo

north

west ☆ east

south

Around the Neighborhood

Mrs. Bunny and her family live in a friendly neighborhood.
There are homes for Mrs. Bunny and her neighbors.
There are shops and a nice playground for the little bunnies.

Mrs. Bunny has many places to go today.
Use the map on page 19 to help Mrs. Bunny find her way.

Get a blue . Make a line where goes.

1. Start at the . Go to the .

2. Start at the . Go to the .

3. Start at the . Go to the .

4. Start at the . Go to the . Then go to the .

5. Start at the . Go to the .

EMC 4130

Mrs. Bunny's Neighborhood maps

FOOD

Store

Carrots

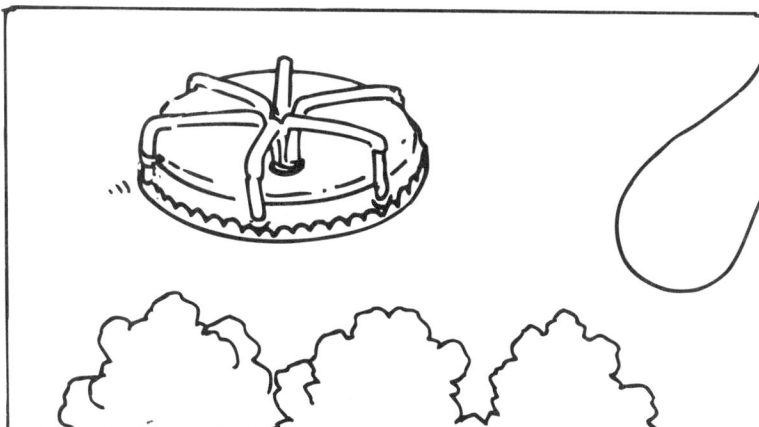
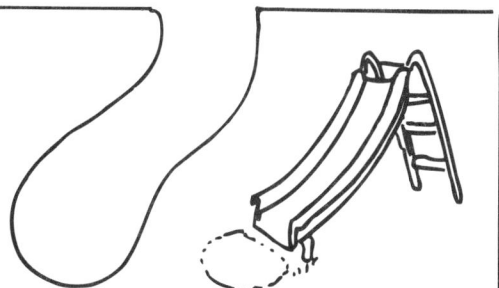

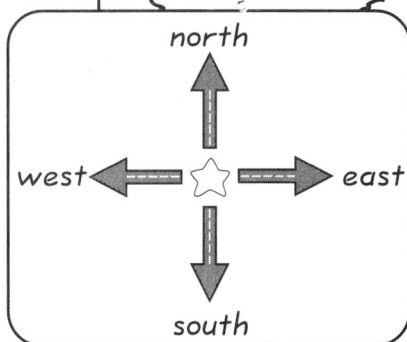

north

west ☆ east

south

The Earth

You live on a planet called Earth.
The Earth is round.
It has land and water.
This is a drawing of the Earth.

EMC 4130

A Globe

This is a globe.

It is a model of the Earth.

The globe shows the land and the water.

Color the globe.

Paste north and south.

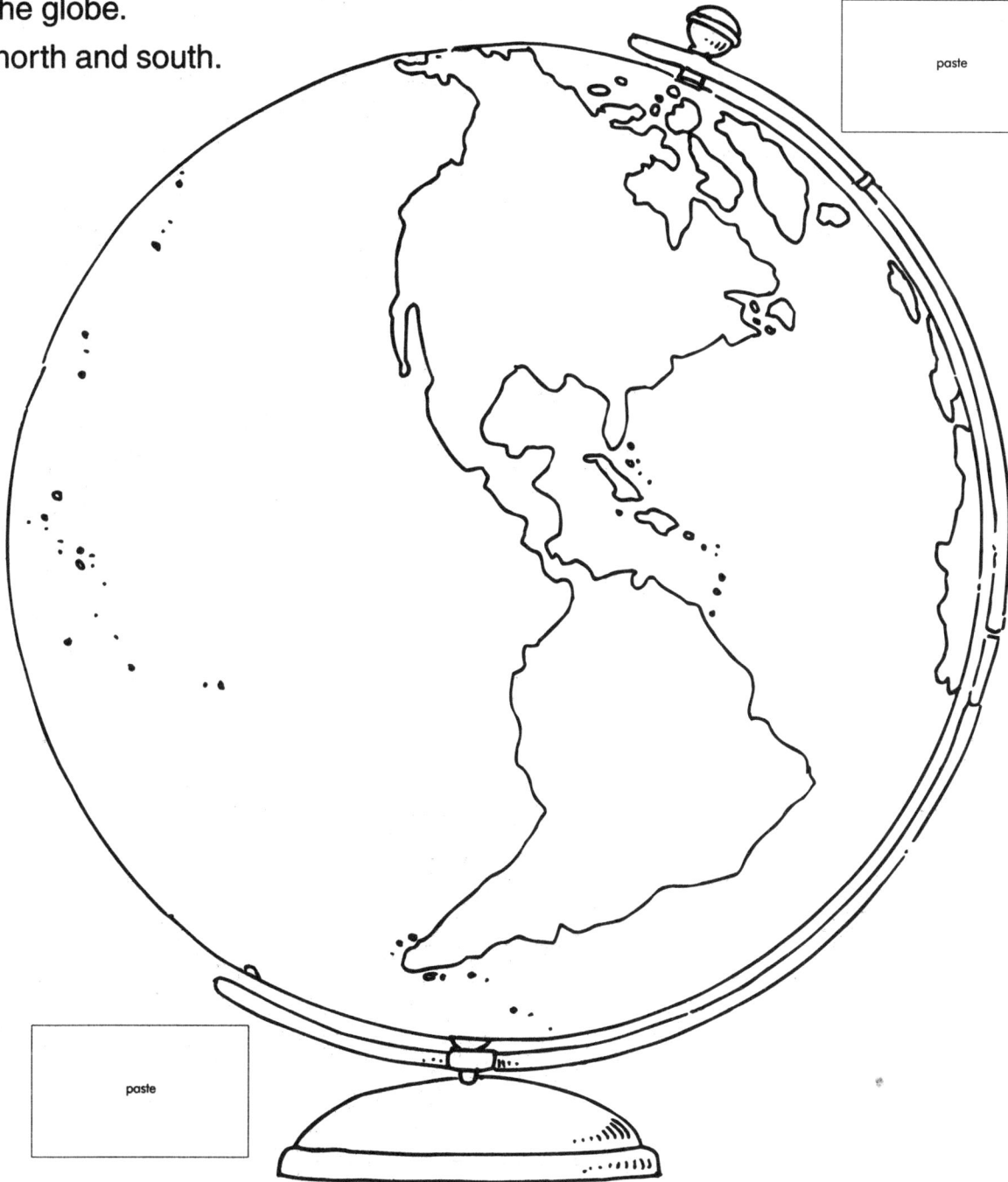

paste

paste

north   south

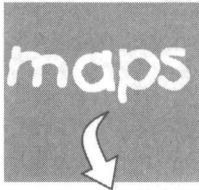

Map

This is a map.

It is flat.

It shows the Earth too.

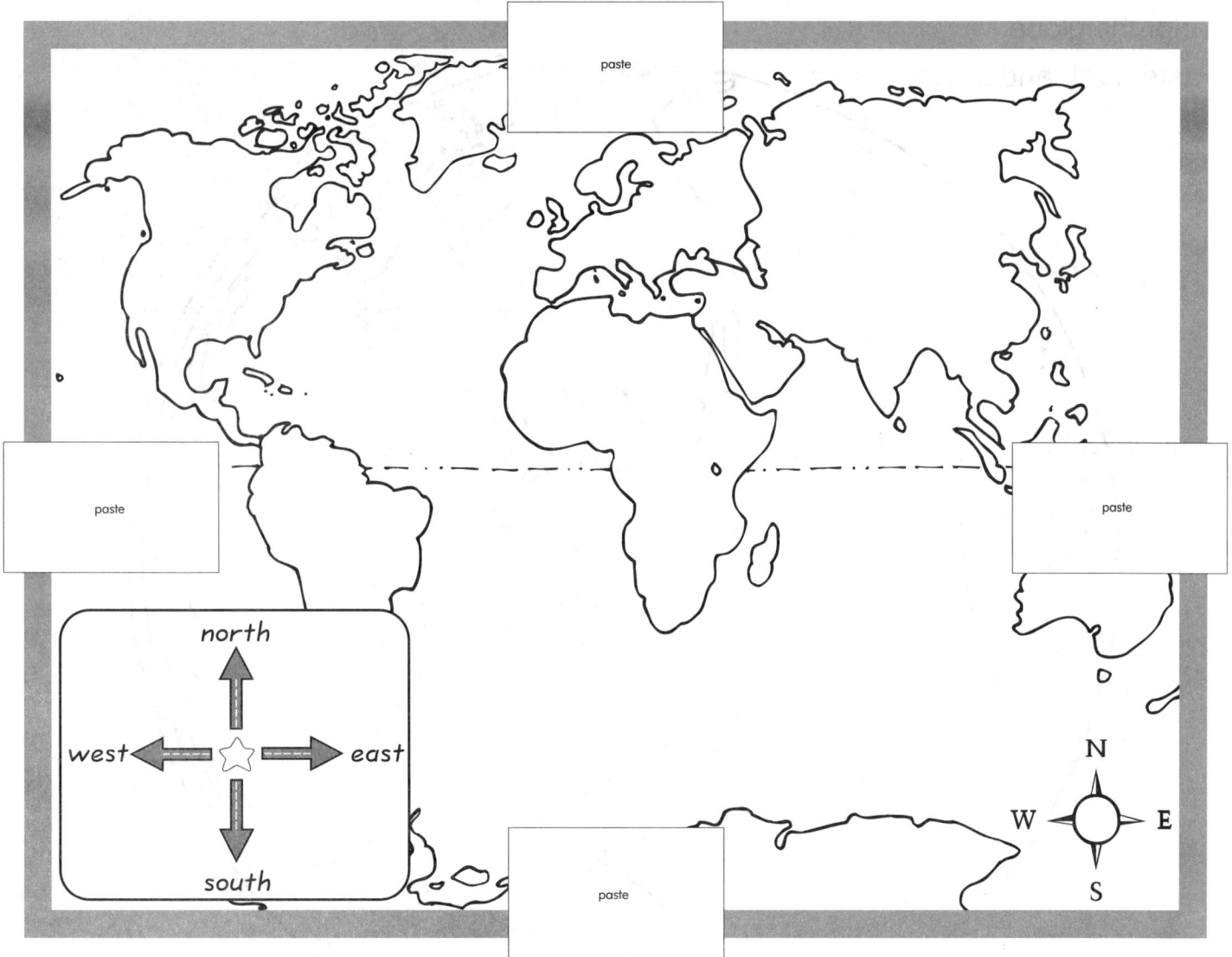

paste

paste

paste

```
           north
             ↑
             |
west ←——— ☆ ———→ east
             |
             ↓
           south
```

paste

N
W ✦ E
S

Color the map.

Paste north, south, east and west.

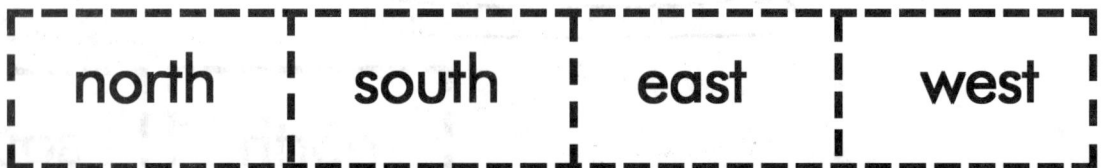

| north | south | east | west |

EMC 4130

Cut out for page 5.

Lunch Table

maps

Parents: Your child is to cut out the pictures on page 23 and paste them on this table just as they

are placed on the map.

Set the Table

5

EMC 4130

Play Rug Map

maps

Parents: Your child is to cut out the pictures on page 25 and paste them on this rug just as they are placed on the map.

My Play Time Rug

Go to page 25.
Cut out the pictures.
Paste them here to
show the toys on
my play rug.

7

EMC 4130

Bedroom Map

maps

Parents: Help your child mark the places where windows and doors belong on the box below. Have your child use the symbols for bedroom furniture on page 27 to make a map of his/her own bedroom.

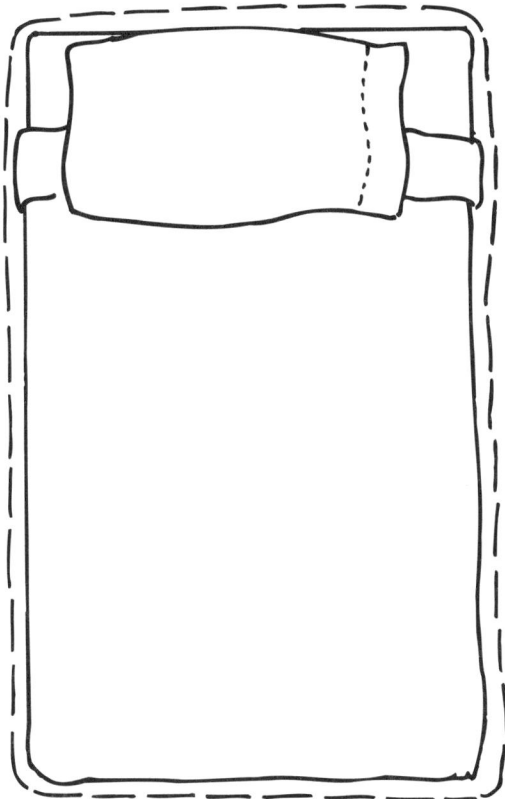

Map Your Bedroom

9

TOYS

28

Little Boy Blue

maps

Little Boy Blue

north

west   east

south

14

Little Bo-Peep

Little Bo-Peep

15